DISCARD

Yukon Elementary Library
Shedeck

Where Is Water?

by Robin Nelson

Lerner Publications Company · Minneapolis

Where is water?

Water is in an ocean.

Water is in a river.

Water is in a **waterfall.**

Rain is made of water.

Clouds are made of water.

Snow is made of water.

Icebergs are made of water.

Water is in a hose.

Water is in a **sprinkler.**

Water is in a swimming pool.

Water is in a **skating rink.**

Water is in a washing machine.

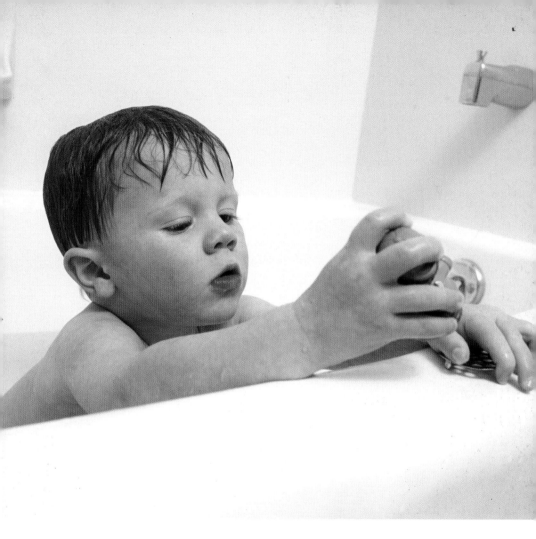

Water is in a bathtub.

Water is in a **fountain**.

Water is everywhere.

Where is water on a map?

How can you find water on a map? Look for the color blue! You can find a lake by looking for a blue area with land all around it. An ocean is a larger blue area. A river is a thin blue line. Sometimes water is shown using other colors. Icebergs are white areas.

Water Facts

- About 80% of the earth's surface is water.

- The tallest waterfall in the world is Angel Falls in Venezuela. It drops 3,212 feet!

- The river that carries the most water in the world is the Amazon River in South America. It carries about 8 trillion gallons of water per day!

- The longest river in the world is the Nile River in Africa. It is 4,132 miles long.

- The deepest lake in the world is Lake Baikal in Siberia. It is 6,365 feet deep. Lake Baikal is also the oldest lake in the world. It is 25 million years old.

- The largest ocean in the world is the Pacific Ocean. It covers 64 million square miles.

- The amount of water on the earth never changes. The water you drink today could have been around when the dinosaurs were alive.

Glossary

 fountain – a stream or jet of water

 icebergs – very large pieces of ice floating in the ocean

 skating rink – an area with a smooth surface for skating

 sprinkler – something that sprays water on grass

 waterfall – a stream of water that falls from a high place

Index

clouds – 7

fountain – 16

hose – 10

icebergs – 9, 19

ocean – 3, 19

skating rink – 13

sprinkler – 11

waterfall – 5

Copyright © 2003 by Robin Nelson

All rights reserved. International copyright secured. No part of this book may be reproduced, stored in a retrieval system, or transmitted in any form or by any means—electronic, mechanical, photocopying, recording, or otherwise—without the prior written permission of Lerner Publishing Group, except for the inclusion of brief quotations in an acknowledged review.

The photographs in this book are reproduced through the courtesy of: US Forest Service, front cover; NASA, p. 2; © Jeff Greenberg, p. 3; © John Rice, p. 4; Minneapolis Public Library, pp. 5, 7, 22 (bottom); © Diane Meyer, pp. 6, 14; © Beth Osthoff/Independent Picture Service, p. 8; National Science Foundation, pp. 9, 22 (second from top); © Larry & Rebecca Javorsky/Photo Agora, p. 10; Rain Bird, pp. 11, 22 (second from bottom); Wisconsin Dells Visitor & Convention Bureau, p. 12; © Heather Robertson/Photo Agora, pp. 13, 22 (middle); Rubberball Productions, p. 15; © A.A.M. Van der Heyden/Independent Picture Service, pp.16, 22 (top), © Joerg Boetel/Photo Agora, p. 17.

Lerner Publications Company
A division of Lerner Publishing Group
241 First Avenue North
Minneapolis, MN 55401 USA

Website address: www.lernerbooks.com

Library of Congress Cataloging-in-Publication Data

Nelson, Robin, 1971–
    Where is water? / by Robin Nelson.
        p.   cm. — (First step nonfiction)
    Summary: Summarizes where water is found on Earth, including lakes, icebergs, and a garden hose, and introduces related concepts such as how to find water on a map and facts such as how high the highest waterfall is.
   ISBN-13: 978–0–8225–4592–7 (lib. bdg. : alk. paper)
   ISBN-10: 0–8225–4592–6 (lib. bdg. : alk. paper)
   1. Water—Juvenile literature. 2. Hydrology—Juvenile literature. [1. Water.] I. Title.
II. Series.
GB662.3 .N47 2003
551.48—dc21   2002007193

Manufactured in the United States of America
2 3 4 5 6 7 – DP – 10 09 08 07 06 05